LAUNCH
VEHICLES

HERITAGE OF THE SPACE RACE

BY MICHAEL LENNICK

For my parents, who had just

the right books on their shelves.

Published by Apogee Books, Box 62034, Burlington,
Ontario, Canada, L7R 4K2, http://www.apogeebooks.com
Tel: 905 637 5737

Printed and bound in Canada

Launch Vehicles -Heritage of the Space Race - by Michael Lennick
Apogee Books Pocket Space Guide #4
ISBN 1-894959-28-0
ISBN-13 978-1-894959-28-5

LAUNCH VEHICLES

Chinese scholars generally ascribe the earliest use of simple rockets to the late thirteenth century AD, though songs written several hundred years earlier offer tantalizing hints of similar technology. The basic secrets wouldn't have been difficult to discern for the thoughtful agrarian; Gunpowder is a blend of charcoal, sulfur and an oxidizer such as potassium nitrate, also known as saltpeter. Since both sulfur and saltpeter sublimate as powder residue through the drying of barnyard feces and urine, the still-warm embers from a household fireplace tossed out the door could easily have provided the charcoal, producing an unexpected lightshow - and an intuitive spark.

There's no way to know just where, when, or in how many places on earth this happened, though evidence of the consequences is plentiful; A few years of harmless fireworks celebrations followed by several more centuries of high-tech bloodshed - cannons, long rifles, handguns and simple hollow-tube war rockets - before a few remarkably prescient humans dreamed up something far more interesting to do with all that kinetic energy.

The Space Age was given practical form near the dawn of the 20th century by four men living in different corners of the world, three of whom never met and, from all accounts, appear to have been unaware of each other's existence. But unlike rocketry, inspiration cannot occur in a vacuum. For the full story we have to go back two earlier generations to meet another man - a frustrated author then working at his brother-in-law's brokerage firm in Paris.

Even as telegraph cables were being laid across the Atlantic,

France was enduring an era of government-sanctioned scientific illiteracy, Louis Napoleon having granted a virtual monopoly to the Catholic orders to run the national school system. At the urging of his radical, soon-to-be-exiled publisher, author Jules Verne set out to underpin his new adventure tales with scientific theory and reason, culminating in the 1865 publication of his novel "From The Earth To The Moon".

The very few space adventures published before Verne's time were fanciful constructs - balloons, sailing ships and even the odd animal-drawn carriage navigating the ether between the heavenly spheres. Verne's work was different. From the heart of the Steam Age the young author set out to establish the practical requirements for a manned voyage to the moon. Although written mainly as a satirical metaphor, Verne's scrupulous research and mathematical verisimilitude (much of it worked out with the aid of his stockbroker brother-in-law) resulted in some profound conclusions. Thus did his three travellers launch from the east coast of Florida on a four-day journey to the moon, whereupon they utilized rockets to propel and steer their vessel home to a climactic ocean splashdown - a little more than a century before that very mission would be flown by the crew of Apollo 11.

Only Verne's method of propulsion left something to be desired. Extrapolating from military state-of-the-art, the author proposed that his venturers tuck into a gigantic bullet-shaped metal projectile, which would then be launched from the largest cannon imaginable. There were two basic problems with this scheme: In

Jules Verne

order to achieve lunar trajectory the vehicle would have had to carry some additional source of thrust that would kick in after the initial cannon lob, so that their projectile might continue to accelerate beyond earth's escape velocity of 17,500 miles per hour all the way to the 25,000 miles per hour speed necessary to reach lunar orbit. Also, and perhaps more significantly, the G-forces from the acceleration of such a launch would have instantly compressed the crew into a gelatinous muck against the rear wall of their vehicle's well-appointed cabin.

An early edition of Jules Verne's *From the Earth to the Moon*

And yet the rest of Verne's speculations were so well reasoned, so compelling, that for the next few decades his novel did what good science fiction has done ever since; it inspired other brilliant minds to start filling in the gaps. As a result, the theories and technology that would ultimately lead to Apollo evolved with near-simultaneity in three very different countries and cultures, each triggered, in whole or part, by Verne's tale, and the need to solve the one untenable technical detail his story relied upon.

THE PIONEERS

In Kaluga, Russia, a shy, backwoods mathematics teacher named Konstantin Tsiolkovsky began a long series of thought exercises inspired by Verne's novel. His private musings and sketches led to a profound series of essays on the practical design of a space-going vessel. In 1883 Tsiolkovsky wrote: "Consider a cask filled with a highly compressed gas. If we

Konstantin Tsiolkovsky

open one of its taps the gas will escape through it in a continuous flow, the elasticity of the gas pushing its particles will also continuously push the cask itself. The result will be a continuous motion of the cask." With this single idea, Tsiolkovsky had grasped a fundamental concept that would continue to elude even some scientists (and most of their benefactors) for decades to come: Exhaust gasses do not require an atmosphere to push against in order to propel a rocket forward. Tsiolkovsky had correctly interpreted Sir Isaac Newton's Third Law of Motion: "For every action there is an equal and opposite reaction." Newton's law also explains why, for example, when we step off a boat onto a dock, the boat tends to move in the opposite direction. Understanding this simple and elegant principle enables us to move from the boat to the shore without ending up in the intervening lake. Tsiolkovsky was the first to understand, or at least express in his writings, that it will also allow us to travel in space, once we've sorted out a few of the minor technical issues.

Born far too soon to ever see his theories tested, the unassuming mathematics teacher lived and died in relative obscurity, appreciated first-hand only by his students and family. Yet his work endured, eventually a young Transylvanian student named Hermann Oberth would independently arrive at similar conclusions in the early 20th century.

When young Hermann turned eleven, his mother gave him a copy of Verne's "From The Earth To The Moon". As he later recalled, he "instantly read it through at least five or six times,

and eventually knew it by heart." By
the age of fourteen, Oberth had
already decided that while gunpowder and expansive gasses
might have their place, only highly combustible liquid fuel
burned under great pressure
would provide the necessary
thrust to escape earth's atmosphere.

Although he had no practical means
to test his "recoil rocket", Oberth   Hermann Oberth
continued to advance and develop
his theories. While still in his teens he realized that the higher the
ratio between the mass of his theoretical rocket and its
propellant load, the faster it would be able to travel. The problem was this: As the rocket burns fuel, its mass (not including
propellant) remains the same, in essence becoming heavier
and heavier in relation to the engine's thrust. Oberth's solution
was to split the vehicle into a series of stages, each containing
its own fuel and rocket engines. His idea was simplicity itself;
As the rocket expends fuel from one set of tanks, those tanks
become dead weight. Jettisoning them and the heavy engines
they fed would enable second-stage engines fueled by their
own fresh tanks to add velocity to the accelerating vehicle far
more efficiently, using as many new sets of tanks and engines
(stages) as needed for the voyage. As Oberth wrote, "The
requirements for stages developed out of these formulas. If
there is a small rocket on top of a big one, and if the big one
is jettisoned and the small one ignited, then their speeds are
added."

While still in his mid-teens, before even deciding what university courses to apply for, Hermann Oberth had defined the fundamental principles that propel every manned space vehicle that's ever left our planet.

Oberth's ideas were advanced beyond even the understanding of those professors assigned to adjudicate them, with the result that, in 1922, his doctoral thesis on rocketry was rejected as nonsense. Oberth later described his reaction: "I refrained from writing another dissertation, thinking to myself: Never mind, I will prove that I am able to become a greater scientist than some of you, even without the title of doctor." A year later, he self- published his theories in a slim volume entitled "Die Rakete Zu Den Planetenraumen" (The Rocket Into Planetary Space). The frequently-republished book was hailed around the world as a work of tremendous scientific importance.

By this time Oberth had emerged as a minor celebrity among physicists and engineers in Germany - in fact he inadvertently established a tradition that continues to this day by becoming the first scientist to advise on a major motion picture. In 1929, director Fritz Lang (of "Metropolis" fame) hired Oberth to provide the necessary technical details for his space travel epic "Die Frau Im Mond" (The Girl In The Moon). The film featured a huge, multi-stage lunar spacecraft that would inspire countless young dreamers for generations to come. In fact, the sequence showing the huge vehicle and its support structures rolling out on tracks from the pro-

Oberth's 1923 book
*The Rocket Into
Planetary Space*

tective assembly building and heading for the water-cushioned launch pad was almost a dress-rehearsal for the Saturn/Apollo lunar launches that would follow forty years later.

A generation of young space enthusiasts were inspired by Lang's film. In Germany, Russia and elsewhere, amateur rocket clubs sprang to life, terrifying and enthralling passersby in equal measure with their loud and frequently explosive attempts to crack the peaceful skies.

Robert Goddard

Around the same time in America, a young physics professor was encountering his own difficulties with a less-than-comprehending press and public. Like his contemporaries in Europe, Massachusetts native Robert Hutchings Goddard had been inspired by Verne's novel. Even as a child, the simple pleasure of lighting a Fourth of July rocket left him wondering whether such unfettered energy might one day be harnessed and put to practical use.

As a young man, Goddard attended a series of lectures by the astronomer Percival Lowell, who had published a vast body of work about canals and other indications of advanced Martian civi-

Percival Lowell

Herbert George Wells

lization he believed he had seen through his 24 inch refractor telescope. Science fiction author H.G. Wells had drawn his own inspiration from Lowell's writings for his 1897 novel "The War Of The Worlds". Wells's work, along with Garrett P. Serviss's less-remembered sequel "Edison's Conquest of Mars", were published in the Boston Post in 1898. These fantastic tales, combined with earlier inspirations, churned in the head of the teenage Robert Goddard one afternoon as he pruned the upper branches of his aunt's cherry tree. As he completed his work and daydreamed about the heavens, Goddard found himself devising a plan that would evolve into his life's work and mission. For the rest of his life Goddard would cite that day, October 19th, 1899, as the anniversary of his career-defining vision. Four years before the first aircraft flew, Goddard determined to build a machine that could travel to the planet Mars.

Years of studying physics led to Goddard's professorship at Clarke University in Worchester, Massachusetts, the institution he would be associated with for the rest of his life. After-hours experiments in the basement at Clarke led to his first patents in 1914, for early liquid fuel and multi-stage designs.

Rocketry research is expensive. Shortly after America entered World War One in 1917, Goddard demonstrated the practical advantages of Newton's Third Law of Motion by launching a small, solid-fuel rocket from a tube attached to a flimsy music stand. The stand didn't even sway, proving the viability of a "recoil-less grenade launcher". Goddard received an army commission to develop such a device, which would feature a shaped charge capable of piercing the armored hull of the new super-weapon, the tank.

Clark University Worcester

Goddard's invention, the Bazooka (named in honor of a then-popular musical instrument it loosely resembled), was accepted into service too late to make an impact in World War One, though it would provide critical infantry support in World War Two and beyond. Goddard was starting to attract attention.

Interested in finding ways to take measurements beyond earth's atmosphere, Charles Abbot, director of the Smithsonian Astrophysical Observatory, arranged for his parent organization to fund Goddard's experiments with a small annual stipend. In 1919 the Smithsonian Institution published Goddard's early theories under the title "A Method of Reaching Extreme Altitudes". The reaction by the public, and the press in particular, astonished Goddard. Articles about "the Moon Man of Massachusetts", along with disturbing reports of loud noises and the occasional fire on his experimental proving grounds (Aunt Effie's farm) dismayed the shy inventor, even as one notorious 1920 New York Times editorial took him to task for seemingly misunderstanding the fundamentals of his craft ("That professor Goddard does not know of the need to have something better than a vacuum against which to react… Of course he only seems to lack the knowledge ladled out daily in high schools…") It's worth noting that The New York Times, that august paper of record, eventually published a retraction and apology for the inaccurate editorial - in 1969, shortly after the flight of Apollo 11. Although far too late to placate the long-deceased Professor Goddard, the Times retraction nonetheless soothed many still-ruffled feathers in the scientific community.

Goddard did not react well to publicity, nor to the inquiries of neighbors or even his fellow experimental physicists. For the rest of his career Goddard would avoid interviews and resist sharing his ideas and discoveries with colleagues, even going so far as to hire workers who would follow his instructions explicitly, often without full knowledge of the grander schemes they were laboring towards.

Goddard's - and the world's - first successful liquid fueled rocket was launched in 1926. It flew for less than three seconds but managed to climb forty-one feet before crashing one-hundred and eighty-four feet downrange. Despite this milestone and others to follow, Goddard's work in Massachusetts was drawing to a close. Endless public scorn and ridicule, not to mention one too many letters from anxious travellers volunteering for any up-coming moon voyages, had led him to search for a more secluded workspace. But the annoying publicity also led to one of his luckiest breaks, when aviator and newly minted hero Charles Lindbergh decided he wanted to meet the "Moon Man of Massachusetts". Lindbergh and Goddard instantly connected, inspiring the influential aviator to bring his friend Harry Guggenheim around for a visit. For the rest of the 1930s, Goddard's work would be financed by a combination of Smithsonian grants and the underwriting of the Guggenheim Foundation, run by Harry's father Daniel. This gave Goddard the freedom at long last to pull up stakes and move his operation to the vast terrain and open skies just outside the town of Roswell, New Mexico.

Robert Goddard and his assistants with Charles Lindbergh
and philanthopist Harry Guggenheim

Goddard continued to work in relative seclusion, gradually refining his designs until it all came together in the P series, built and flown between 1939 and 1941. The P-type rockets burned gasoline and liquid oxygen in a highly advanced combustion chamber. Per Newton's Third Law, a rocket's speed is a direct function of the amount of mass you can throw away behind you, and the velocity with which you can throw it. Goddard's earlier rockets used pressurized gas to feed the combustion chamber. The P-type employed high-speed turbine pumps designed and built by Goddard to force huge quantities of fuel and oxidizer into the chamber and out the nozzle. The rocket was steered by metal vanes that reached directly into the hot exhaust gasses, under the guidance of what is probably his greatest contribution to the field, a three-axis gyroscope that, in one form or another, would spin at the heart of

A late model Goddard rocket showing the exhaust steering vanes

every serious rocket or missile launched from that day to this.

Goddard wasn't out to set records or impress anyone beyond his own toughest critic - himself. His handmade rockets were rebuilt after every crash, improving incrementally with each new fix. No Goddard rocket ever reached space nor flew more than a few miles. His exhaust thrust never exceeded 1,000 pounds. Yet his desert breakthroughs in the 1930s and early 40s solved the most fundamental problems, establishing patents and designs that underpin all modern rocketry.

Robert Goddard poses with one of his late model rockets at his facility in Roswell, New Mexico

ROBERT H. GODDARD'S EXPERIMENTAL ROCKET LAUNCHES

- First successful liquid-fuel flight: March 16th, 1926

- Altitude reached: 41'

- Distance downrange: 184'

- Average speed: 60 mph

- Fuels: Gasoline and liquid oxygen

- Flight duration: 2.6 seconds

- Peak performance flight: October 19th, 1941

- Rocket designation: P-C

- Altitude reached: 9,000'

NOTEWORTHY GODDARD FIRSTS (PARTIAL LIST)

- First U.S. patent for multi-stage rocket - 1914

- First flight of liquid fuel rocket, March 16,1926

- First flight of scientific payload (barometer and camera) 1929, Auburn, Massachusetts

- First developed pumps suitable for rocket fuels

- First use of guidance vanes in rocket exhaust - 1932, New Mexico

- First use of gyroscopic control - 1932, New Mexico

- First flight of rocket motor pivoted on gimbals under gyroscopic control - 1937

THE MAN WHO AIMED AT THE STARS

Around the same time on the other side of the planet, another young rocketeer was getting ready to make his own impact on history. Equally inspired by Verne's tale, Wernher von Braun had emerged as one of the brightest members of Hermann Oberth's civilian rocket club, the VfR (Verein fur Raumschiffart.)

The Treaty of Versailles that had ended World War One was very specific; Germany was forbidden to develop most known weapons of war. However the treaty's architects lacked the foresight to include any mention of rockets. In the early 1930s, Army Colonel Walter Dornberger, intrigued by the prospects of using small, solid-fuel rockets as battlefield weapons, visited the VfR. Wernher von Braun, the club's star designer, was soon offered an army commission, along with full university tuition to complete his studies. Von Braun had little interest in warfare but knew his ultimate goal - to design and build a rocket capable of carrying a man into space -

Walter Dornberger late in life as chief scientist at Bell Aerospace

An advertisement placed by VfR member Rudolf Nebel, encouraging people to join the Rocket club

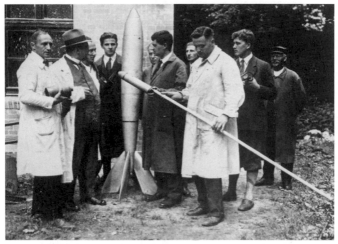

Left to right: Rudolf Nebel, Franz Ritter, unknown, Kurt Heinisch, unknown, Hermann Oberth, unknown, Klaus Riedel,
Wernher von Braun, unknown. Riedel holds an early version of model for the minimum rocket 'Mirak'.

would not be financed through an amateur rocket club. The young engineer enthusiastically accepted Dornberger's offer, bringing much of the club along with him. Less than six months later, Adolph Hitler assumed power. One of the Nazi regime's first moves was to disband the VfR. All further rocketry experiments would now be under military control.

By the early 1930s von Braun's team outgrew their limited facilities at Kummersdorf, eventually relocating to the remote and picturesque setting of Peenemünde on the Baltic coast. Here von Braun would have the test beds, the wind tunnels and fabrication shops to fulfill his dreams, for he had never lost sight of his true goal

Wernher von Braun in
the 1960's

- that of building a rocket capable of exploring space. A gradual progression of vehicles eventually culminated in Peenemünde's masterpiece, the A-4, a 46 foot, 26,000 pound ballistic missile with an alcohol and liquid oxygen-burning turbine-fed engine that could generate a launch thrust of 56,000 pounds, increasing to 160,000 pounds at full speed. After two failed attempts the A-4 was successfully launched on October 3rd, 1942. Rising to a peak altitude of 50 miles, it eventually crashed into the sea 120 miles downrange. This was the weapon the Führer was waiting for. Armed with a one-ton warhead and a range of nearly 250 miles, the A-4 was renamed Vengeance Two (V-2) by Hitler. In the aftermath of the 1943 bombing of Peenemünde by British forces, the V-2 project was placed under the command of SS Commandant Heinrich Himmler, and moved to an underground facility in central Germany known as the Mittelwerke (Metal Works). Prisoners from nearby Concentration Camp Dora were brought into the tunnels to churn out the vast numbers of rockets Hitler demanded.

A World War 2 era aerial reconnaissance photograph of the Peenemünde launch facility.

On September 8th, 1944, the first operational V-2 was launched against Paris. Over the next six months some 3,000 V-2s were fired against civilian populations in London, Antwerp and Paris, causing thousands of casualties. Although the V-2 created terror by striking its targets without warning at more than three times the speed of sound, it was ultimately a failure as a military weapon; A profoundly expensive way to deliver a single one-ton warhead, the terror

A trio of V2/A4 missiles at Peenemünde awaiting launch.

weapon wasted essential personnel and resources without significantly changing the outcome of the war. In fact, far more people died building the rockets than were killed through their subsequent use. It has been suggested that tens of thousands of slave laborers left the Mittelwerke only as smoke.

Von Braun continued his work at Peenemünde, far from the death tunnels of the Mittelwerke. History remains unclear on his direct involvement with or knowledge of the brutal conditions on the production side, but it is generally acknowledged that he was deeply upset by the deadly toll his missiles were taking in England and Europe. When congratulated on an early, successful flight against London, he's reported to have replied, "Yes, but we're hitting the wrong planet."

The entrance to the notorious Mittelwerke and inside (below)

Neither this attitude nor von Braun's constant talk of moon rockets and space expeditions impressed Himmler's SS, and in 1944 he was arrested and charged with treason. It took the direct intervention of der Führer (under the guidance of Albert Speer, himself a space enthusiast) to save von Braun's life and return him to his post at Peenemünde. For the remainder of the war Himmler's spies kept the scientist under close scrutiny.

By early May of 1945 Hitler was dead, and the major Allied powers were scouring Germany in search of the rocket scientists and their priceless heritage of unfired V-2 rockets. Von Braun's team voted almost unanimously to surrender to the Americans, sending Wernher's brother Magnus out to meet an American patrol and arrange terms. A carefully coordinated effort known as Operation Paperclip managed to smuggle out about 100 top scientists and engineers, along with enough parts to build an equivalent number of V-2s. Von Braun and his team were relocated to a facility in White Sands, New

Mexico. Calling themselves "prisoners of peace", they would spend the next several years refining the V-2 for the nascent American space effort. Meanwhile, the Soviets captured and reactivated the production facility at the Mittelwerke, in part by enslaving several of the original workers who were hiding in the area. The fledgling American rocketry program, eventually under the leadership of von Braun, and the Russian effort under Soviet designer Sergei Korolev, relied heavily on German V-2 technology to get off the ground. For the remainder of his remarkable career, von Braun, the man whose work would eventually land Americans on the moon, played down his role in the Nazi war machine, though not his importance to the history of rocketry. In 1959 Columbia Pictures released a bio-pic of the rocket pioneer entitled "I Aim At The Stars". It didn't take long for commentators to add "...and hit London."

The legendary group of scientists rounded up in Germany under *Operation Paperclip* at the end of World War 2.

A V2 on its way to wreak havoc during World War 2

A-4 (V-2) SHORT RANGE BALLISTIC MISSILE

- Height: 46 feet

- Weight : 26,000 pounds

- Fuel: Alcohol

- Oxidizer: Liquid oxygen

- Engine thrust: 56,000 - 160,000 pounds

- Maximum speed: Mach 5

- Maximum range: 250 miles

- Capacity: 1 ton high-explosive warhead

The United States managed to bring many V2 components back to White Sands Missile Range where they would be tested. The Soviets were not so fortunate but managed to replicate the missile with their R-1 (above left) The American's added an upper stage called a WAC-Corporal which set a new altitude record (above right)

Looking like something out of a science fiction movie, a dramatic shot taken at White Sands in the late 1940's of a captured V2 being prepared for launch.

THE BLIP THAT CHANGED THE WORLD

By the early 1950s Wernher von Braun, now leading the American Army's rocketry program from his new base at the Redstone Missile Arsenal in Huntsville, Alabama, was spending much of his time promoting the idea of launching an artificial satellite into earth orbit. He met great resistance from many in the Eisenhower administration, who were fearful of challenging the Soviet Union to a competition for the high ground of space. A breakthrough of sorts came when permission was granted to launch a satellite as part of the upcoming International Geophysical Year, actually an 18-month span extending from July of 1957 through the end of 1958. Von Braun proposed a joint Army-Navy project code-named Orbiter, which would launch a Jet Propulsion Laboratory-built satellite atop the Army's Jupiter-C missile, an upgraded version of von Braun's Redstone missile (which was itself an upgraded version of his World War II V-2.) The administration set up a

competition between von Braun's proposal, the Navy's Project Vanguard, and an Air Force plan to launch a satellite aboard its still-unflown Atlas missile. In mid-1955 the choice was made to go with the Navy's Project Vanguard. A 72 foot long, three-stage liquid-fuel rocket would launch a 6.4 inch, 3.25 pound aluminum tracking satellite. Von Braun and the Jet Propulsion Laboratory were deeply disappointed, but the choice was as much political as scientific; Vanguard was a civilian-managed booster, which satisfied Eisenhower's

The US Navy's Project Vanguard launch vehicle

desire to distance the military from the IGY project.

VANGUARD LAUNCH VEHICLE

- First launch: 1956

- Length of service: 1956 - 1959

- Height: 72 feet

- Diameter: 3 feet, 9 inches

- Number of stages: 3

- Fuel: Kerosene RP-1

- Oxidizer: liquid oxygen

- Engine thrust: 27,000 pounds

- Maximum speed: Mach 25

- Number of attempted launches: 14

- Number of successful launches: 3

The world changed radically on October 4th, 1957, when the Soviet Union's Chief Designer, Sergei Korolev, launched a basketball-sized, 183 pound sphere dubbed Sputnik 1 into low-earth orbit aboard his R-7 five-stage rocket. The launch vehicle performed flawlessly, releasing the world's first artificial satellite into an elliptical orbit that circled the earth every 98 minutes, emitting a beeping telemetry signal that could be picked up by any backyard observer equipped with a short-wave radio. The new moon's polar orbit guaranteed it a worldwide audience, as it drifted through the night skies of every nation on earth - a breathtaking accomplishment to some, and a dire warning to others.

The unexpected launch of Sputnik shocked the world - and the administration. Schedules were compressed in order to

The fireball that was to have
been America's first satellite.
Vanguard fails December 1957

get Vanguard off the pad
before the end of 1957. On
December 6th, all was ready;
The world's first live televised
launch started off perfectly; As
the countdown reached zero
the engine fired. The rocket
lifted a few feet off the pad,
then settled back down upon
itself in a devastating, TV
screen-engulfing fireball. The
headlines shouted out endless
variations of "Kaputnik!" in
the first major embarrassment
handed to U.S. pride by
Korolev and his Soviet rocket
team. It would not be the last.

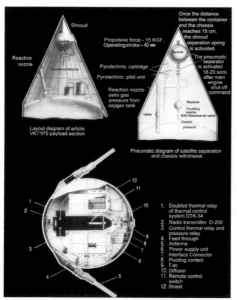

The world's first
artificial satellite,
called Sputnik.

Korolev's R-7 rocket
which launched Sputnik

R-7 LONG RANGE BALLISTIC MISSILE

- First launch: 1957

- Height: 90 feet

- Weight: 27 tons (unfueled), 280 tons (fueled)

- Number of stages: 2 (without warhead or upper stages)

- Fuel: Kerosene T-1

- Oxidizer: liquid oxygen

- Maximum speed: Mach 5

- Maximum range: 5,000 miles

- Stage 1 burn time: 104

-130 seconds from launch

- Stage 1 propulsion: 1 x four-chamber main engine RD-107 (8D74), 2 x one-chamber steering engines

- Stage 2 burn time: 285-320 seconds

- Stage 2 propulsion: 1 x four-chamber main engine RD-108 (8D75), 1 x four-chamber steering engine

- Storage time (fueled): 30 days

- Readiness time: 2 hours

Fortunately, von Braun and the JPL satellite Explorer One were standing by. On January 31st, 1958, in the wake of a second successful Soviet launch, they finally got their chance.

The Jupiter-C was basically a Redstone with tanks extended to provide longer thrust. The liquid-fuel booster was topped by three solid-propellant upper stages arranged in a unique configuration within a spinning cylinder. The second stage was an outer ring of eleven scaled-down JPL-built Sergeant rockets, each generating 16,500 pounds thrust over a burn time of 6.5 seconds. The third stage was an inner cluster of three 5,400 pound thrust Sergeant rockets. The fourth stage contained the Explorer vehicle itself, propelled by a single 5,400 pound thrust Sergeant rocket. To stabilize this upper stage stack, the entire cylinder was spun up on the pad, rotating between 450 and 750 rpm. At T minus zero the Jupiter's stage one Rocketdyne A-7 engine fired, its 83,000 pounds of thrust burning for two and a half minutes before releasing the upper-stage cluster.

Von Braun and James van Allen celebrate their success with a full size mock-up of the Explorer 1 satellite.

Explorer One achieved an elliptical orbit with a perigee of 224 miles, an apogee of 1,575 miles and an orbital period of 115 minutes. The 31 pound satellite soon fulfilled its mandate when an instrument designed by JPL's Dr. James Van Allen discovered intense bands of radiation circling the earth. The Van Allen belts were deemed to be one of the prime discoveries of the International Geophysics Year. Explorer One went on to become the longest-orbiting artificial satellite in history. In fact it's still up there now, and is expected to continue its journey for at least another thousand years.

JUPITER C INTERMEDIATE RANGE BALLISTIC MISSILE

- First launch: 1955
- Length of service: 1955-1964
- Height: 60 feet, 1 inch
- Diameter: 8 feet, 9 inches
- Contractors: Chrysler (missile body), Rocketdyne (1st stage liquid-fuel engine)
- Number of stages: 1 (could be adapted for upper stages)
- Fuel: Kerosene RP-1
- Oxidizer: liquid oxygen
- Engine thrust: 150,000 pounds
- Maximum speed: Mach 8
- Maximum range: 1,976 miles

A Jupiter C launch vehicle carrying America's first satellite vents in the night air.

Unlike the top-secret Russian space program which was conducted far from prying eyes and television cameras, the American space effort was a very public affair. President Dwight D. Eisenhower had served as the Supreme Commander of the Allied Forces during World War Two, and was well aware of the dangers inherent when military secrecy combines with the "Skies the limit" avarice of free market industrial capitalism. Eisenhower warned of the dangers of what he termed the Military-Industrial Complex the day before leaving office in 1961. His administration had been forced to feed the ravenous beast throughout the burgeoning Cold War that would define his two-term presidency, so when it came time to establish a government bureaucracy to oversee the American answer to Sputnik, Ike refused to follow the recommendations of his military advisors. Instead, in 1958 he converted the National Advisory Committee for Aeronautics (NACA), the fifty-year-old civilian agency tasked with the exploration of flight, into the National Aeronautics and Space Administration (NASA). The new civilian agency was given a mandate to investigate and establish an American manned presence in space. Less than three years later Eisenhower's successor, John F. Kennedy, would hand NASA an even larger assignment - the moon.

Under Eisenhower's directives, the new agency would do most of its work in the public eye, wide open to media scrutiny. Fully funded by tax dollars but run and staffed by civilians, the agency would be expected to put out bids to American industry for whatever it couldn't develop or build in-house. All new materials, processes and inventions developed for NASA programs would belong to the American public except for those fully devised by private contractors, and even they were to be made available for license to American industry and the public at large. Even the footage shot of rocket launches and tests, failures as well as successes, was to be considered public domain - a boon to historians and documentary filmmakers to this day.

Although the technology required for later manned missions such as Apollo and the Space Shuttle would be developed specifically for those purposes, the launch vehicles for early

satellite and manned flights were adapted from existing ballistic missile programs. These would not be limited to the government's own medium-range Redstone missile that Wernher
von Braun had adapted from his original V2 design (right down
to the graphite vanes that steered the vehicle by physically
deflecting the exhaust - very similar to Goddard's original
design), but would also include Convair's Atlas D and the
Martin company's Titan II missiles, both capable of hurtling a
warhead to any point on earth, or placing a spacecraft in low-
earth orbit.

Throughout NASA's heady early days, the technologies and
techniques for launching humans into space and retrieving
them safely were being invented on a day-by-day basis. Project
Mercury, initially run out of Langley Field in Virginia, soon followed Christopher Kraft and his Space Task Group team to
their new home and launch facility at Cape Canaveral, Florida
- a site chosen not only for its (mostly) favorable climate and
existing Naval Air Station facilities, but also for its flight-path
out over the Atlantic Ocean, well away from civilian populations when the inevitable launch disasters occurred. The cape
offered another distinct advantage; Launching an orbital vehicle near the equator in the same direction as the earth's rota-
tion was like gifting your vehicle with a thousand-mile-an-hour
tailwind.

Christopher Kraft came from aircraft flight-test, and so the
early straight-up-and-down sub-orbital lobs planned for
Project Mercury were not too extreme a leap. It was the later
missions - orbital flights over parts of the earth unequipped
for even primitive telegraph communications - that gave him
the most worry. The thought of a journey to the moon and
back was beyond his wildest imaginings when the NASA
Mercury program began. Initial concerns, such as establishing a
real-time worldwide radio and telemetry link for the upcoming orbital missions, were enough to keep Kraft and his crews
busy full time. Kraft set out to invent a means of overseeing

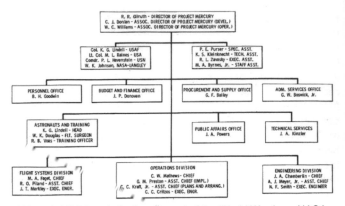

Although NASA technically grew out of the World War I era NACA, the real business of space exploration began with the formation of the Space Task Group in Langley Virginia. A small group of scientists and engineers from around the world,

the launch and flight of his highly complex vehicles by inventing something he called "Mission Control" - a room full of flight controllers, mission specialists and experts from the various civilian contractors who could react to ever-changing circumstances on a moment's notice, sorting out the problems and sending any necessary instructions upstairs to the orbiting astronauts. Mission Control evolved over the years, eventually splitting its responsibilities between Launch Control (which took place in a massive room overlooking the launch pad three miles distant at the cape) and Mission Control, which would be housed at the Johnson Space Center in Houston, Texas, and assumed command the moment the vehicle cleared the launch tower.

Kraft's teams trained intensely, ran endless simulations, and prided themselves on the ability to solve any problem that might arise. Certainly the events that took place subsequent to the unprecedented explosion on board the CSM of Apollo 13 en route to the moon bear witness to the skill and excellence of Kraft's teams and training (Flight Director Gene

Kranz has called those four days "NASA's finest hour"), but the early evolution of Mission Control's methodologies are equally telling. During a late-stage unmanned test flight of the Mercury Redstone combination (with the Mercury astronauts in attendance), something went wrong. The countdown reached zero but the rocket failed to ignite. Instead, the Mercury spacecraft's solid-fuel escape tower fired its own jettison engines, departing the silent rocket at tremendous velocity. Neither astronauts nor flight controllers knew quite what happened or what to do about it, as the Mercury spacecraft let off an extravaganza of parachutes, dye markers and other debris, all of which now dangled down the side of the vehicle ("Just like Mrs. Murphy's bloomers.", as Mercury astronaut Wally Schirra later described the sight.) Concern filled the air in Mission Control; A fully-fueled and pressurized rocket was sitting on the pad, capable of exploding at any moment. When one member of the team suggested getting a high-powered rifle and punching a few holes in the rocket's tanks to relieve pressure, Kraft got on the loop and suggested that it was time to establish "the first rule of Mission Control: If you don't know what to do, don't do anything!" And that's just what they did. Eventually the pressurized tanks outgassed and the rocket's systems safed themselves, ready for another try. As Gene Kranz later suggested, Kraft's Law has gone on to be enshrined as a pretty good rule of thumb, "both for the business of Mission Control and pretty much every other aspect of life."

The Mercury-Redstone configuration went on to reliably launch both chimpanzees and Mercury astronauts into sub-orbital lobs. Although the first manned flight was scheduled for March of 1961, von Braun and his team were nervous; The January flight of Ham the chimp had not gone well. The Redstone burned for longer than anticipated, creating problems for the spacecraft and its occupant. Although the engineers felt they had the problem under control, they wanted one more chimpanzee flight to make sure. Their wish was granted, and the first manned Mercury/Redstone launch was moved back to May. On April 12th, 1961, Sergei Korolev launched cosmonaut Yuri Gagarin on a single-orbit mission aboard an R-7 rocket. And the world stood in wonder once again.

On May 5th, 1961, Alan Shepard became America's first man in space when a Redstone missile launched his spacecraft, Freedom 7, on a journey that would carry him to 104 miles altitude and 250 miles downrange, ending sixteen minutes past T-zero in an ocean splashdown. Three weeks later, on May 25th, President Kennedy addressed Congress, setting his nation on course to land a man on the moon before the end of the decade. Newly-appointed NASA Chief Administrator James Webb called for budget proposals while his engineers and flight directors worked to push the program into high gear. On July 25th, 1961, astronaut Gus Grissom flew a nearly-identical mission to Shepard's in a spacecraft he'd named Liberty Bell 7, aboard a Redstone booster. Additional sub-orbital flights were scheduled, but in the wake of Cosmonaut Gherman Titov's August/1961 seventeen-orbit triumph, the decision was made to forego all remaining Redstone sub-orbital flights and pro-

ceed directly to the orbital missions, using the then far less reliable Atlas D booster.

MERCURY-REDSTONE LAUNCH VEHICLE

- Height: 83 feet
- Diameter: 5 feet, 10 inches
- Number of stages: 1
- Fuel: Kerosene RP-1
- Oxidizer: liquid oxygen
- Engine thrust: 78,000 pounds
- Maximum speed: Mach 8
- Maximum range: 200 miles (Ballistic missile version)

AN ATLAS-SIZED BELLY ACHE

It was the combination of open media coverage and the difficult process of adapting launch vehicles that were never
intended to be man-rated that resulted in some very legiti-
mate concerns among the public - and the newly-appointed
Mercury astronauts. Correspondents such as Walter
Cronkite, on his weekly series "CBS Reports" and "The 20th
Century", would regularly show footage of massive missiles
exploding in the skies over southern Florida, as technicians
tackled the near-impossible task of modifying these incredibly
powerful machines to carry their precious human cargo to
space without incident and with 100 percent reliability. It was
these well-publicized, spectacularly telegenic failures that
resulted in the public attitude, best summed up by author Tom
Wolfe in his book "The Right Stuff", of "Our rockets always
blow up." At the time, they did. The launch of a rocket is best
viewed as a barely-controlled explosion, in which vast energies
are triggered, contained temporarily, and directed along a single axis opposite the vehicle's intended direction of travel.
Spectacular, even tragic mistakes light up the sky only when
those dynamic forces find more than one way to escape the
vehicle. Thus the success, power and useful load-carrying ability of any launch vehicle is the direct consequence of a delicate
engineering balance between the dynamic force potential of
the quantity and nature of the propulsive fuel, and the strength
- and thus weight - of the vehicle's many containers, fittings and
external skin. With limited resources in the wake of World
War Two, the Soviet Union was unable to direct its industrial
base to create the lightweight metal alloys to benefit such a
weight/thrust venture. Instead, Korolev's engineers designed
and built huge, multi-thruster multi-stage vehicles like the massive R-7, overpowered of necessity just to lift its own weight.
The design and engineering were so robust that variations on
those early models were still in use at the turn of the millen-
nium. American industry, with multiple strategic missiles to
build, transport, store and maintain, took a very different
approach.

The Atlas ballistic missile featured a lightweight aluminum outer shell so thin it needed to be supported on the pad with pressurized gas, prior to fueling, to prevent the empty vehicle from collapsing under its own weight. As the fuel was consumed on the climb to orbit the thin hull would actually begin to flex under the dynamic forces, making for a very rough ride for Mercury astronauts Glenn, Carpenter, Schirra and Cooper. The remarkable power of the 82.5 foot tall Atlas D missile resulted from a unique combination of five liquid-fuel engines in a one-and-a-half-stage configuration - a central sustainer engine, two large boosters, and two small vernier engines delivering a total thrust of 360,000 pounds. At T plus 130 seconds the lower skirt containing the two booster engines was jettisoned, leaving the rocket climbing to orbit on the steerable sustainer engine and the two side-mounted verniers. At T plus 300 seconds the remaining engines shut down, having delivered the rocket to a sustainable low-earth orbit at a velocity of approximately 17,500 mph.

The Atlas remains notorious for the large number of highly-photogenic failures during its conversion from a warhead launcher into America's prime manned launch vehicle. Numerous test launches, many attended by the Mercury astronauts themselves, would end ignominiously in a blazing fireball filling the blue skies over Canaveral. The problem was eventually traced to the join between the Mercury spacecraft and the upper stage of the missile. As the spacecraft was far bigger and heavier than the nuclear warhead the missile had been designed to carry, the offset load combined with the flexing of the thin-skinned vehicle set up a resonance not unlike a flexing diving board, causing the assembly to tear itself apart. The problem was eventually solved through the addition of a "belly band" - a strap of metal wrapped around the upper portion of the vehicle at the point of maximum dynamic stress. Although confident that their vehicle was now safe for manned flight, the astronauts were understandably apprehensive when John Glenn finally climbed aboard for the Mercury program's first orbital mission. As of that February morning in 1962, three of the previous five Mercury/Atlas test launches had failed spectacularly. Fortunately the care, precision and diligence observed by the rocket engineers, as well as Christopher Kraft's flight controllers and Guenter Wendt's pad team, paid off for Project Mercury, as the four manned Mercury/Atlas launches went off without a hitch.

MERCURY-ATLAS LAUNCH VEHICLE

Mercury-Atlas

- Height: 82 feet, 6 inches
- Diameter: 10 feet
- Number of stages: 1.5
- Contractor: Convair Corp.
- Fuel: Kerosene RP-1
- Oxidizer: liquid oxygen
- Engine thrust: 360,000 pounds
- Maximum speed: Mach 25
- Maximum range: 10,360 miles (Ballistic missile version)

PROJECT GEMINI AND THE BIG RED CLOUD

Nuclear-tipped missiles serve a very different mandate than do manned launch vehicles. The precision, power and safety required of a manned vehicle made liquid-fueled rocket engines the best possible way to go. The thrust of a liquid-fueled engine can be throttled, both for efficiency and comfort, and to safely guide the vehicle through the aerodynamic extremes encountered as it passes through "Max-Q", the region where the density of the air is at its highest relative to the rocket's velocity. A liquid fuel rocket can be instantly shut down in case of fire or other emergency simply by slamming valves closed or shutting off the turbine pumps. Liquid fuel, in short, is controllable - far more so than solid rocket boosters, although they too have their place, as launchers for fast-response nuclear weapons, and to augment first-stage power on both unmanned and, more recently, manned flights.

Solid boosters offer simplicity and phenomenal power, but come with a huge trade-off in safety. Basically a solid fuel rocket is nothing more than a long hollow tube typically filled with a mixture of ammonium perchlorate (the oxidizer) and fine aluminum powder (the fuel), held together in a base of PBAN or HTPB (highly flammable rubber compounds.) Poured into molds wet, the mixture dries, lining the walls of the cylinder, leaving a hollow space running down the long vertical axis; cylindrical near the top and star-shaped towards the base to sculpt the exhaust. Once the igniter atop the stack is sparked the fuel fires, burning furiously along the full length of the motor from the hollow core out towards the cylinder walls, venting its powerful exhaust out the open base. Once started, a solid-fuel rocket cannot be shut down. This is why a rocket stack that uses a combination of liquid and solid engines, such as NASA's Space Shuttle, will fire its liquid-fuel engines first and make sure they're burning properly, before igniting the strap-on solids and leaping off the pad.

Project Gemini was a hybrid - a series of two-man missions designed to develop and practise techniques needed for the lunar missions to come. While Project Apollo's launch vehicles and spacecraft were under construction, a scaled-up version of the Mercury spacecraft was put into production. Large enough to hold two astronauts, the Gemini spacecraft would feature new computers, fuel cells to power extended missions, orbital maneuvering rocket motors for rendezvous and dock-ing, newly-designed spacesuits, tethers and hatches to allow for extra-vehicular activity (EVA), and the flight dynamic con-trols necessary to pilot the craft accurately through a high-speed controlled re-entry - all essential mission requirements if Project Apollo was to succeed.

Unfortunately, the Atlas missile wouldn't be powerful enough to lift the larger Gemini spacecraft to orbit. With no other dedicated launch vehicle available, the world's largest and most powerful nuclear-tipped missile was pressed into service: the Air Force's two stage Titan II. Like the Atlas missile, the Titan II had not been designed with the safety or comfort of a human passenger in mind, which could make for a very unpleasant ride. One of the worst, and potentially most hazardous, side effects of early liquid-fuel designs was called POGO, a phenomenon that approached near-disastrous levels on Gemini missions. In a liquid-fueled vehicle, vibrations during liftoff and ascent can cause the flow of propellants in feed lines to oscillate, which could lead to minute oscillations in engine thrust. The result was a sort of push-pull effect along the vehicle's longitudinal axis, or direction of travel; hence the name POGO, for it truly felt to the increasingly-rattled human passengers like they were riding a pogo stick to orbit. Even worse, the thrust oscillations could increase the vehicle's overall vibration, leading to even greater thrust oscillations in a progressive feedback loop. These conditions wouldn't have caused much concern for the nuclear warheads the vehicle was originally designed to carry, but they made for a very uncomfortable and quite alarming ride for Project Gemini's human passengers. The Titan II was a 102-foot ballistic missile designed to be launched against enemy targets on the other side of the world from hardened underground silos. While a manned space mission would allow a pad crew the time to assemble, pressurize and test a liquid fuel rocket prior to launch, the instant response demanded of a nuclear-delivery system, combined with the cramped conditions of an underground launch silo, rendered unsuitable standard super-chilled liquid hydrogen and liquid oxygen fuels, along with their massive support apparatus and crews. Instead, the Titan would use room-temperature hypergolic fuels. Far more efficient and reliable, hypergolic fuels are chemicals that react violently the instant they're brought into contact with each other. At launch, valves open and a great wash of monomethyl and dimethyl hydrazine (the fuel) and nitrogen tetroxide (the oxidizer) pour into the bell-shaped thrust chambers, where they instantly ignite, generating, in the case of the Titan II, a thrust of 216,000 pounds per chamber. No spark or other ignition source is required. After a burn of 155 seconds the first stage is released, and the single, stage two, 100,000 pound thrust engine is ignited to carry the Gemini spacecraft into orbit.

The problems stemmed from the fact that hypergolic fuels are incredibly toxic, both to most standard materials (it was quite a trick coming up with storage systems and high-speed pumps and turbines that could handle exposure to these caustic chemicals), and to any human unlucky enough to breath their fumes. (A reality that continues to this day. The maneuvering thrusters on the Space Shuttle also use hypergolic fuels, which must be purged before returning astronauts can safely disembark from their vehicle on the runway.)

The scene around the Cape Canaveral launch pad is always one of great risk during a launch, but that hazard was more than doubled during Project Gemini, as the launch vehicle's engines instantly created a huge plume of unbelievably dangerous vapor, known to astronauts and engineers alike as the B.F.R.C., or "Big (expletive) Red Cloud".

On December 12, 1965, Gemini 6 stood ready to launch. The mission had failed to get off the pad once already, scrubbed when the Atlas missile that was to place an Agena Target Vehicle into orbit for rendezvous and docking practice blew up shortly after liftoff. Now astronauts Wally Schirra and Tom Stafford were waiting out the countdown on a new mission; Instead of rendezvousing with an Agena, they would race to meet and fly formation alongside Frank Borman and Jim Lovell, already orbiting aboard Gemini 7.

The countdown proceeded smoothly to T-zero, the Titan II's engines fired, burned for a couple of seconds while computers waited for thrust to stabilize before releasing the rocket's hold-down clamps, then abruptly shut down. (It was later determined that one of the booster's electrical connections had slipped from its socket, triggering an automatic, computer-controlled abort.) On board Gemini 6, Commander Schirra was faced with the most difficult decision of his career - whether or not to follow outlined procedure and pull the ejection lanyard. Much more than the mission was at stake; Early designs had determined that the Gemini spacecraft could avoid the expense, risk and indignity of a mid-ocean splash-down by landing on dry land under a Rogallo wing - an early prototype of today's popular parasail sportcraft. As the space-craft would come in horizontally, standard pilot's ejection seats were considered mandatory, enabling designers to do away with the solid-fuel escape rocket that had topped the Mercury vehicle on the pad. Unfortunately the parasail landing tech-nique was scrapped early in development. Gemini would come in for a standard splashdown, but it was too late to re-design the launch escape system. This meant if Schirra twisted the abort handle, instead of being carried up and away from the pad by an escape rocket, he and Stafford would eject straight out into the already massive toxic plume of rocket exhaust. (Gemini pilots, no fans of the compromised escape system, referred to the plan as "Committing suicide to avoid being killed.") Consummate astronaut that he was, Wally Schirra kept his cool. When flight controllers assured him they were studying the situation, he piped back with, "Okay, we're just sit-ting up here breathing."

Gemini 6 was successfully shut down, the crew disembarked safely. Three days later the repaired vehicle was launched into history. Schirra and Stafford's mission went off without a hitch, pushing NASA a critical step closer to the moon. Despite their inherent risks, hypergolic fuels would continue to play essen-tial, even life-saving roles in future NASA missions.

A Martin Titan II
ballistic missile
modified and
carrying the two-man
Gemini spacecraft

GEMINI-TITAN LAUNCH VEHICLE (TITAN II)

- Height: 102 feet
- Diameter: 10 feet
- Number of stages: 2
- Contractor: Martin Company (later Martin Marietta)
- Fuel: Monomethyl and dimethyl hydrazine
- Oxidizer: Nitrogen tetroxide
- Engine thrust: 2 X 216,000 lbs
- Maximum speed: Mach 25
- Maximum range: 9,325 miles (Ballistic missile version)

TEN TIMES FASTER THAN A RIFLE BULLET

Even as NASA's Space Task Group was making their plans to utilize military vehicles like the Redstone and Atlas missiles for Project Mercury, Wernher von Braun and his team in Huntsville were designing the future. A long-term space program aiming for more than one or two astronauts on temporary assignment in low-earth orbit would require far more lifting capacity than could be provided by existing or planned military launchers. As with any new transportation enterprise, there would need to be a whole new infrastructure. Von Braun anticipated the need for new support facilities on earth, as well as a huge manned orbiting structure - a wheel or space station, slowly spinning to provide gravity for its permanent staff and crew. These incredible designs became very familiar to a space-obsessed young audience in the 1950s, primarily through the promotional efforts of von Braun himself. Not willing to risk the same short-sighted rejection his grand ideas had received a decade earlier, von Braun went public. Books written with his old VfR colleague Willy Ley (and illustrated by the great space artist Chesley Bonestell), magazine articles in LOOK and COLLIERS, "Tomorrowland" television specials on Walt Disney's popular weekly television series (von Braun even had a hand in designing the "Tomorrowland" section of the new "Disneyland" theme park), all served to keep our potential near-future ablaze in the minds of a young and starry-eyed post-war generation. Hollywood, too, was paying attention; Feature films from George Pal's 1950s epics "Destination Moon" and "Conquest of Space" (both featuring Bonestell designs) to Stanley Kubrick and Arthur C. Clarke's late-60s masterpiece "2001: A Space Odyssey" were directly influenced by von Braun's orbiting space wheel and other futuristic visions.

Von Braun and his team at Huntsville's Redstone Arsenal, now renamed the George C. Marshall Spaceflight Center, were already hard at work designing the hardware for a round-trip to the moon. Based on designs they'd worked out for the Army Ballistic Missile Agency (ABMA) back in 1957, their goal was a rocket capable of generating a first-stage thrust of one point five million pounds to lift three astronauts and tons of lunar exploration and survival gear into orbit, then accelerate them to the speed necessary to break earth orbit and head for the moon - twenty-five thousand miles an hour; Ten times faster than a rifle bullet.

The first Saturn launch vehicle would be based on existing Jupiter and Redstone technology, using a unique "clustered" design. The original idea was to group eight Redstone missiles around a central Jupiter missile core in order to utilize the combined lifting power of all nine rockets firing in unison. Eventually the decision was made to use the Jupiter and Redstone rockets only as fuel tanks, four each for RP-1

Kerosene and liquid oxygen, which would then feed eight Rocketdyne H-1 engines mounted in two clusters of four engines each, for a total thrust of one million, three-hundred and thirty-thousand pounds. The two-stage Saturn C-1 was test-launched in 1961 with the inert second stage filled with water to provide proper balance and weight.

The Saturn C-1 would eventually evolve into the very successful Saturn 1B launch vehicle, used to lift the Apollo Command and Service Modules (without the Lunar Module) into earth orbit. The vehicle would prove critical to the success of Apollo 7, the first orbital test of the new Apollo spacecraft, and would serve as the launch vehicle for the crews of all three Skylab missions, as well as the joint American/Soviet mission known as the Apollo/Soyuz Test Project (ASTP) on July 15th, 1975.

SATURN C-1 LAUNCH VEHICLE

- Height: 164 feet
- Diameter: 21 ft, 6 inches
- Number of stages: 2
(second stage inert)
- Fuel: Kerosene RP-1
- Oxidizer: Liquid oxygen
- Engine config (stage 1)
8 Rocketdyne H-1 engines
- Engine thrust (total):
1,330,000 pounds
- Maximum speed:
Mach 25

A Saturn I is shown to scale side by side with a Mercury Redstone.

SATURN 1B
LAUNCH VEHICLE

- Height: 224 feet
- Diameter: 21 ft, 6 inches
- Number of stages: 2
- Fuel: Stage I: Kerosene RP-1
- Oxidizer: Liquid oxygen
- Engine config (stage 1) 8 Rocketdyne H-1 engines, thrust (total): 1,640,000 lbs
- Stage 2: Single Rocketdyne J-2 engine, Fuel: Liquid hydrogen / liquid oxygen, thrust: 225,000 lbs

The Saturn IB.

As far back as the early 1950s, long before President Kennedy committed America to a lunar program, Wernher von Braun had already mapped out his strategy. The earliest lunar landing scheme was known as Direct Ascent, and would require a massive multi-stage rocket dubbed the "Nova", which would feature eight 1.5 million pound thrust F-1 engines driving its first stage alone. Nova would shed multiple stages as it accelerated an Atlas-sized upper-stage Command Vehicle that would eventually land on the moon, then lift off and return its crew to earth. It was an audacious scheme, and would have looked for all the world like the classic lunar landings described in such 1940s science fiction literature as Robert A. Heinlein's "Rocketship Gallileo" and 1950s films like "Destination Moon". (In those days, as SF author Robert J. Sawyer has said, all rockets were direct ascent vehicles, landing tail fin first, "as God and Heinlein intended.") Unfortunately, Nova was not to be. The reasons were many, though the most critical might have also been the most mundane; The Saturn assembly facility, a refurbished W.W.II munitions plant in Michoud, Louisiana, had a ceiling height of just over thirty-six feet, and so would barely be able to accommodate the more realistic if still massive Saturn Five and its thirty-three foot diameter first stage, which would eventually house five F-1 engines.

NASA engineer John Houbolt had a different idea, and although it took him a very long time to get NASA management to pay attention, his far-more-practical plan would end up carrying the day. Houbolt called his approach "Lunar Orbit Rendezvous (LOR). It called for a single launch, carrying both an Apollo Command vehicle and a much smaller landing craft that would carry two astronauts to the lunar surface while the third waited in orbit, then be abandoned once the crew had reunited. Houbolt's plan was far more cost-effective in that no earth orbit infrastructure was required, and the entire mission could be carried out with only a single Saturn Five launch.

The F-1 engine, the largest rocket motor ever built, stood nearly nineteen feet tall and weighed a little under ten tons. The diameter of its exhaust bell was nearly fourteen feet across - large enough to engulf a good-sized car. The most efficient burn in a liquid fuel engine is derived from a mixture of hydrogen and oxygen, but there's a hefty price to be paid; The fuel-to-oxidizer ratio is set by the elements' atomic numbers. For every gram of hydrogen (the lightest element with an atomic number of one, or one proton per atomic nucleus) you need eight grams of oxygen (atomic number eight). That's an enormous amount of otherwise inert oxidizer to lift from a standing start. Thus, for the initial leg of its journey the Saturn Five would burn a far less exotic fuel, kerosene. During its two-and-a-half-minute lifespan each F-1 consumed two-thousand tons of kerosene and oxygen at a rate of three tons per second, giving the massive booster an overall fuel efficiency rating of just under five inches to the gallon.

The Saturn Five first stage was originally designed to hold four F-1 engines, but reports were coming in from NASA that the rapidly-evolving Apollo program was already having trouble meeting its weight limits. Four F-1s would be just barely adequate to lift the Command, Service and Lunar modules to orbit along with the massive rocket's own weight. As gear and experiments were added to the Apollo package von Braun grew increasingly apprehensive, eventually deciding to squeeze a fifth F-1 into the limited space at the center of the engine cluster. The new engine wouldn't be able to gimble with the others as there was no room for those mechanisms, but its extra 1.5 million pounds of thrust would make all the difference. The added engine eventually made possible the successful "J" missions, the scientific expeditions which capped the Apollo program. These longer-duration missions' added launch weight included the new Lunar Rover, heavier scientific payloads and sample return programs, as well as supplies for a three-day lunar lay-over. By the time the J missions flew, those five blazing F-1s would provide the safety margin needed to lift the six-million pound vehicle slowly off the launch pad.

The Saturn Five did indeed end up with its own outsized support structures here on earth. Von Braun and his engineers realized the size and weight of the vehicle made the lengthwise assembly techniques of earlier Saturns impractical - wires and delicate piping could easily be crushed. The safest approach would be to assemble the giant rocket vertically, using massive cranes to lift each stage and lower it gently onto the stack. This necessitated the construction of NASA's Vehicle Assembly Building (VAB), at the time the largest building (by internal volume) ever conceived. The 50 story VAB contained four "tall bays", each large enough to construct and house a fully-assembled Saturn Five with Apollo spacecraft and pad support tower and launch base structures in place. The entire assembly would be gently rolled out to pad 39, some three miles distant, on another extraordinary device, the crawler/transporter - a diesel-powered, tractor-treaded platform as large as a baseball infield. The crawler carried its fragile multi-million pound cargo at the breakneck speed of one mile an hour along a specially-designed stone roadway as wide as a four-lane interstate highway (including medians.) The crawler's platform could adjust its angle to compensate for the sloping terrain as it nudged the rocket up and onto the elevated launch pad. It was said of this ride that an engineer perched atop the gantry's highest platform could rest his coffee on the railing, secure that not a drop would be spilled on the slow voyage to the pad.

As of the first decade of the 21st century, the Saturn Five rocket still retains its bragging rights as the largest, most powerful machine ever constructed. Television and film cameras may have reduced the scale of a Saturn Five liftoff to something that could fit in a living room, but no recording from that era can adequately convey the sheer power, the shock and awe of the event. Those few individuals this author has spoken with whose government service led them to witness both, often equated the Saturn Five launch to a nuclear test in the Nevada desert, though with a far more upbeat ending. The comparison is apt; The amount of raw power ready to be unleashed in a Saturn Five at launch is indeed equivalent to that of a small nuclear warhead; Were it all to go at once the impact on surrounding territory would be eerily similar. Even a controlled and gradual release through those massive F-1 engine bells left its mark. Viewers and journalists stationed more than three miles away from pad 39 were astonished by the physical effects of the launch. Journalist Walter Cronkite, America's unofficial guide to the space program throughout the 1950s and 60s, expressed great concern during an early Saturn test launch that the CBS News structure was shaking apart, as the huge bay windows flexed wildly and ceiling tiles fell on Cronkite and his guest, "2001: A Space Odyssey" author Arthur C. Clarke. The world would thrill to those unprecedented Saturn Five launches, both manned and otherwise, each slowly climbing into the blue Florida skies, flashing their offset white and black checker paint jobs - an homage to the fictional moon rocket in Fritz Lang's "Die Frau Im Mond" that had so influenced the young von Braun, and had graced every vehicle his team launched since their earliest A-4s in Peenemünde.

SATURN V LAUNCH VEHICLE

- Height: 363 feet
- Maximum diameter: 33 feet
- Number of stages: 3
- Contractors: Stage 1: Boeing, Stage 2: North American Aviation, Stage 3: Douglas Aircraft
- Fuel: Stage 1: Kerosene RP-1, Stages 2 & 3: Liquid hydrogen
- Oxidizer: Liquid oxygen
- Engines
- stage 1: Five Rocketdyne F-1 engines, total thrust 7,500,000 lbs
- stage 2: Five Rocketdyne J-2 engines total thrust 1,000,000 lbs
- stage 3: Single Rocketdyne J-2 engine total thrust: 225,000 lbs

The Saturn V

The Saturn Five took a remarkably long time to fly its own length. Unlike the Space Shuttle which, thanks to its solid rocket boosters, clears the tower in less than three seconds, the Saturn Five took a full ten seconds to achieve that critical milestone, having reached the stately speed of just under sixty miles an hour. More than two-thirds of the rocket's total weight was fuel, most of which would be used to carry the vehicle to low earth orbit.

The stage one engines had the most work to do, gradually lifting the equivalent weight of several buildings off the pad and raising it to an altitude of 41 miles before exhausting their fuel supply at the two point five minute mark. Saturn's on-board computer sensed the S1 tank depletion, shutting down the fuel pumps and blowing a series of explosive bolts to cut the stage loose. Solid-fuel retro rockets fired to slow the first stage and help separate it from the ascending rocket. At the same time, four solid-fuel 21,000 pound thrust ullage rockets mounted on the interstage section would fire to kick the second stage forward, assisting in the separation, and also forcing the second stage liquid fuel to settle at the bottom of the tanks. (This innovation was the result of one of the many back-of-the-envelope engineering solutions von Braun was known for. The combined thrust of the ullage engines far exceeded the power of the Redstone rocket that lifted Alan Shepard to space in 1961.) Once the interstage section falls away the second stage engines would fire. Built to operate in the near-vacuum of the upper stratosphere, the second stage's five J-2 engines were far more efficient than the F-1s, burning a combination of liquid hydrogen and liquid oxygen to develop a variable thrust of around 200,000 pounds each, for a second stage mean thrust of around one million pounds. The second stage fired for about six and a half minutes, bringing the rocket to the brink of orbital velocity before it, too, fell away. The Saturn Five's third stage contained the Apollo Command and Service Modules the Lunar Lander, and a single, re-startable J-2 engine, which would burn for an additional two and a half minutes to push the rocket into a near-circular parking orbit at a velocity of 17,500 miles per hour. After a single earth orbit to verify all systems, the stage three J-2 would be re-ignited for a six and a half minute burn known as a Trans-Lunar Injection, or TLI. This would push the spacecraft to a speed of nearly 25,000 miles per hour and place it in a much wider orbit - one that, after four days of travel, would intercept the orbit of the moon. Once this burn was accomplished the stage three J-2's work was done. Following a much-rehearsed maneuver, the Command Module Pilot would separate the CSM from the third stage, pitch over to face the now-exposed Lunar Module stored above the stage's fuel tanks, and dock with the Lunar craft, withdrawing it from its housing within the now-useless third stage. Both the CSM/LM stack and the exhausted Saturn third stage would continue on course for the moon, the stage crashing into a pre-designated spot on the lunar surface even as Apollo arrived tail-first, firing the CSM's big main engine to decelerate the vehicle into lunar orbit.

Nine Apollo crews flew that epic journey between 1968 and 1972, and for those who were able to stop and explore it was here, at the farthest reaches to date of humanity's personal journey, that the toxic simplicity of their CSM's big hypergolic engine would shine. Once locked into lunar orbit, the astronauts' main engine was their only ticket home. There was no room for the standard NASA redundancy at the business end of the CSM. A failure of complicated ignition circuitry would doom the crew to circle the moon forever, a constant, bitter reminder to future generations of school kids and amateur astronomers of the limits of human engineering. Fortunately NASA's designers, engineers and contractors made the right choices many millions of times over, as they gradually turned ancient fiction into practical reality. At that critical moment towards the end of every Apollo mission, the Command Module Pilot pushed the button that opened mechanical valves to let volatile fuels collide in the combustion chamber, sending the crews hurling through the darkness towards the oceanic splashdown first experienced in fiction by Jules Verne's courageous explorers more than a century earlier. Wernher von Braun, along with NASA's thousands of designers, engineers and outside contractors, had delivered an Apollo program that pushed humans gloriously towards the stars, and changed the night sky forever.

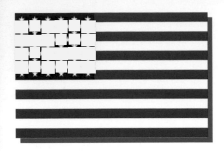

USA

The United States was the second nation to officially enter the space launch field, having been beat out by only two months by the former Soviet Union (USSR). The US has put more launch vehicle types into space than any other nation, but is only second in terms of the actual number of launches.

Vanguard

The US Navy Vanguard project began in 1955, in direct competition with the Redstone and Atlas launch vehicles. Three Vanguard rockets were launched between December 1957 and September 1959 (plus eight launch failures). The first successful Vanguard launch put America's second satellite, Vanguard 1, into orbit.

The Thor-Able (above) was a three-stage rocket using the Thor IRBM as a core booster. It was introduced in 1958 and was later modified to carry small satellites into orbit. Three Thor Able I rockets were launched between August 1958 and November 1958. The Thor-Able I had a fourth stage, and was designed to send the Pioneer spacecraft to a lunar flyby.

The Jupiter C (right) was a multi-stage rocket based on the Redstone, and similar to the Juno I. Six Jupiter C rockets were launched between February 1958 and October 1958.

A direct descendant of the Jupiter C program, Juno (left) was the first non-military launch vehicle, and launched the first US Explorer satellite. The Juno program was eventually reassigned to NASA and renamed Saturn. 10 Juno II rockets were launched between December 1958 and May 1961.

The Mercury-Redstone launch vehicle (right) was a modified US Army Redstone missile. Two Mercury-Redstone launches – both in 1961 – were America's first manned space flights, the suborbital flights of Alan Shepard and Gus Grissom.

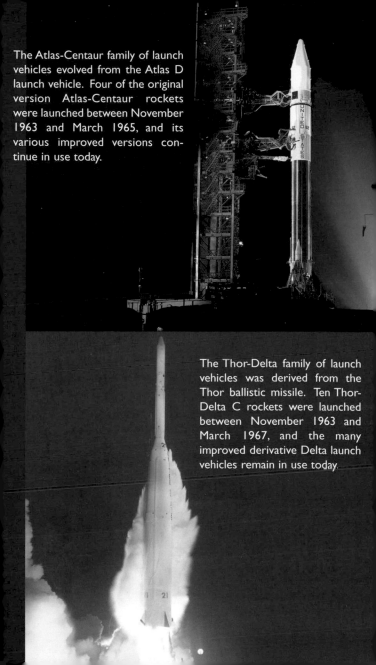

The Atlas-Centaur family of launch vehicles evolved from the Atlas D launch vehicle. Four of the original version Atlas-Centaur rockets were launched between November 1963 and March 1965, and its various improved versions continue in use today.

The Thor-Delta family of launch vehicles was derived from the Thor ballistic missile. Ten Thor-Delta C rockets were launched between November 1963 and March 1967, and the many improved derivative Delta launch vehicles remain in use today.

The Mercury-Atlas launch vehicle was a modified Atlas D rocket. 14 Atlas D rockets were launched between April 1961 and July 1967.

The Atlas-Agena family of rockets used Atlas first stages and Agena second stages. The Atlas-Agena D (called SLV, Standard Launch Vehicle) was used to launch the Gemini target vehicles. 48 Atlas SLV-3 Agena D rockets were launched between August 1964 and November 1967. Shown is the Gemini 11 target Atlas-Agena.

The Titan II GLV (Gemini Launch Vehicle) was a modified version of the Titan II intercontinental ballistic missile used for the Gemini Project. 11 Titan II GLV rockets were launched between April 1964 and November 1966. Shown is the Gemini 4 Titan II GLV.

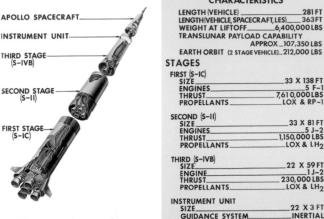

SATURN V LAUNCH VEHICLE

CHARACTERISTICS

APOLLO SPACECRAFT

INSTRUMENT UNIT

THIRD STAGE
(S–IVB)

SECOND STAGE
(S–II)

FIRST STAGE
(S–IC)

LENGTH (VEHICLE) .... 281 FT
LENGTH (VEHICLE, SPACECRAFT, LES) ..... 363 FT
WEIGHT AT LIFTOFF .... 6,400,000 LBS
TRANSLUNAR PAYLOAD CAPABILITY
           APPROX .... 107,350 LBS
EARTH ORBIT (2 STAGE VEHICLE) ... 212,000 LBS

STAGES

FIRST (S–IC)
  SIZE .... 33 X 138 FT
  ENGINES 5 F–1
  THRUST .... 7,610,000 LBS
  PROPELLANTS .... LOX & RP–1

SECOND (S–II)
  SIZE .... 33 X 81 FT
  ENGINES 5 J–2
  THRUST .... 1,150,000 LBS
  PROPELLANTS .... LOX & LH_2

THIRD (S–IVB)
  SIZE .... 22 X 59 FT
  ENGINE 1 J–2
  THRUST .... 230,000 LBS
  PROPELLANTS .... LOX & LH_2

INSTRUMENT UNIT
  SIZE .... 22 X 3 FT
  GUIDANCE SYSTEM .... INERTIAL

MSFC–71–IND 1223M

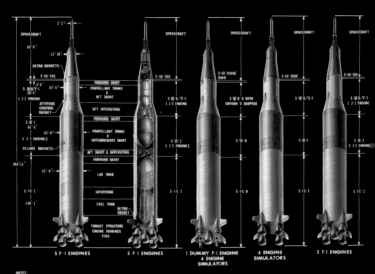

SATURN V CONFIGURATION (R–ME LABORATORY GUIDE)

SA-501	SA-502	SA-500D	SA-500F	SA-503
FLIGHT VEHICLE	FLIGHT VEHICLE	DYNAMIC TEST	FACILITIES CHECKOUT VEHICLE	FLIGHT VEHICLE
5 F-1 ENGINES	5 F-1 ENGINES	1 DUMMY F-1 ENGINE 4 ENGINE SIMULATORS	4 ENGINE SIMULATORS	5 F-1 ENGINES

NOTES:
1. DIMENSIONS SHOWN ARE APPROXIMATE.
2. VEHICLE NUMBERING APPLIES TO LAUNCH VEHICLE ONLY. SPACECRAFT ARE SHOWN FOR REFERENCE PURPOSES.

SATURN I LAUNCH SUMMARY

RESEARCH AND DEVELOPMENT FLIGHTS

SA-1
1. LAUNCHED-OCT. 27, 1961
2. S-I STAGE PROPULSION SYSTEM SATISFACTORY

SA-2
1. LAUNCHED-APR. 25, 1962
2. PROJECT HIGHWATER RELEASED 22,900 GAL. H₂O INTO IONOSPHER

SA-3
1. LAUNCHED-NOV. 16, 1962
2. 2ND PHASE PROJ HIGHWATER
3. FULL PROPELLANT LOADING

SA-4
1. LAUNCHED-MAR. 28, 1963
2. ENGINE OUT CAPABILITY DEMONSTRATED

SA-5
1. FIRST BLOCK II LAUNCHED-JAN. 29, 1964
2. FIRST LIVE S-IV STAGE AND INSTRUMENT UNIT

SA-6
1. LAUNCHED-MAY 28, 1964
2. FIRST ACTIVE GUIDANCE FLIGHT
3. FIRST FLIGHT APOLLO BOILERPLATE AND LES
4. ENGINE OUT (UNPLANNED)

OPERATIONAL FLIGHTS

SA-7
1. LAUNCHED-SEPT. 18, 1964
2. COMPLETELY ACTIVE ST-124 GUIDANCE

SA-9
1. LAUNCHED-FEB. 16, 1965
2. FIRST PEGASUS (METEOROID TECHNOLOGY SATELLITE) ORBITED
3. FIRST UNPRESSURIZED INSTRUMENT UNIT

SA-8
1. LAUNCHED-MAY 25, 1965
2. ORBITED SECOND PEGASUS SATELLITE

SA-10
1. LAUNCHED-JULY 30,1965
2. ORBITED THIRD PEGASUS SATELLITE
3. COMPLETED SATURN I PROGRAM

I-RM-D IND B1048 G

An early view of the Apollo Saturn V Vehicle Assembly Building (VAB) under construction. The building is 525 feet high and covers eight acres of ground and encompasses almost 130 million cubic feet of space. The steel structure was driven 170 feet into bedrock and used 123 miles of steel or 60,000 tons. A million square feet of siding were applied later.

The Saturn IB was an improved version of the Saturn I launch vehicle, used for the early Apollo flights, Skylab and the Apollo-Soyuz Test Project. Six Saturn IB rockets were launched between January 1968 and July 1975.

The Titan III launch vehicles were basically a Titan II plus a third stage. The Titan IIIC was distinguished from its predecessors by the addition of two large strap-on solid rocket boosters. 36 Titan IIIC rockets were launched between June 1965 and March 1982.

The three-stage Saturn V was the largest of the Saturn launch vehicles and was responsible for putting men on the Moon. It is the most powerful rocket ever successfully launched. 13 Saturn V rockets were launched between November 1967 and May 1973 (15 were built).

The Titan IIID launch vehicle was a two-stage version of the Titan IIIC. The third stage was eliminated to make room for larger payloads. 22 Titan IIID rockets were launched between June 1971 and November 1982.

The Titan IIIE Centaur had a Centaur D third stage to provide capability for launching large NASA scientific payloads. Seven Titan IIIE rockets were launched between February 1974 and September 1977.

The Delta II launch vehicle (above) has existed in many configurations. Shown is the Delta 2914 which launched the Westar 1 satellite. The various Delta II models use varying numbers of strap-on solid boosters. 33 Delta 2914 rockets were launched between April 1974 and February 1988.

The Delta L, M, and N launch vehicles, referred to as the Long Tank Deltas (LTD), were the same except for the third stage. The Delta M-6 (right) had a Star third stage motor and a total of six Castor 2 strap-on boosters. 13 Thor Delta M rockets were launched between September 1968 and March 1971.

Cassini's seven-year journey to Saturn began with the liftoff of a Titan IVB/Centaur carrying the orbiter and its attached Huygens probe. Launch occurred on October 15, 1997 from Launch Complex 40 on Cape Canaveral Air Station.

A remote camera captures a close-up view of a Space Shuttle Main Engine during a test firing at the John C. Stennis Space Center in Hancock County, Mississippi.

The Space Shuttle Atlantis lifts off from Kennedy Space Center. 113 Space Shuttle stacks were launched between April 1981 and January 2003. The loss of Space Shuttle Columbia and the STS-107 crew in January 2003 temporarily grounded the US manned space program. The Shuttle program resumed in 2005.

A more powerful version of the Titan I, the Titan II SLV (Standard Launch Vehicle) burns storable hypergolic propellants and can be readied for launch in under 60 seconds. 13 Titan II SLV rockets were launched between Sep. 1988 and Oct. 2003.

The Scout family of small, low-cost rockets was used to launch lightweight satellites and perform high-altitude research. The Scout G-1 (below) was a four-stage vehicle only 21.9m high and 1m in diameter.

The Pioneer missions used an Atlas-Centaur booster with an added third stage to achieve a launch velocity of over 51,500 km (32,000 miles) per hour.

Cassini's seven-year journey to Saturn began with the liftoff of a Titan IVB/Centaur carrying the orbiter and its attached Huygens probe. Launch occurred on October 15, 1997 from Launch Complex 40 on Cape Canaveral Air Station.

The Atlas I launch vehicle has a very thin light-weight single wall of stainless steel (kept rigid by internal fuel pressure) to reduce weight, which increased its range. The Atlas I was also the first instance of using additional boosters that detach and fall away after burn-out to save weight. 11 Atlas I rockets were launched between July 1990 and April 1997.

A recent version of the Delta II launch vehicle is the 7000 series. Shown is a Delta 7925. The 7000 series uses improved strap-on solid boosters (graphite epoxy motors). 34 Delta 7925 rockets were launched between November 1990 and November 1997.

The Atlas II and Atlas IIA have powerful Centaur stages providing greater payload capability. The Atlas IIAS-Centaur (shown here) includes four Castor 4A strap-on solid rocket boosters (the first Atlas to have strap-ons). 27 Atlas IIAS rockets were launched between December 1993 and February 2004.

The Titan IV was developed for military launches and was first used in 1989. A number of versions, with increasing performance, now exist. Shown is a Titan 401B / Centaur vehicle. Seven Titan 401B / Centaur rockets were launched between October 1997 and September 2003.

Pegasus is a launch vehicle that is dropped from an aircraft and then fires its rocket to carry a payload into space. There are several versions of increasing payload capacity. NASA's X-43A tests were performed with the X-43A attached to a Pegasus HXLV dropped from an Air Force B-52.

The Athena 1 and Athena 2 (shown here) are solid-propellant launch vehicles for delivering small payloads to low-Earth, geostationary and interplanetary orbits. An Athena 3 is also being introduced. NASA has included the Athena 2 in its launch services contract with the Delta and the Atlas. Three Athena 2 rockets were launched between January 1998 and September 1999.

The Minotaur launch vehicle is a modified Minuteman ICBM. Its upper two stages are from the Pegasus XL booster. Two Minotaur rockets, one each in January 2000 and July 2000 have been launched.

The Atlas III uses a single liquid-fueled NPO Energomash RD-180 engine (the first use of a Russian-made engine in a US launch vehicle). The second stage is either one or two engines on the Centaur upper stage (Atlas 3A and Atlas 3B, respectively). The Atlas 3B is shown here. Three Atlas 3B rockets were launched between February 2002 and December 2003.

The Delta IV launch vehicle, using a single liquid oxygen / liquid hydrogen RS68 engine, is capable of replacing the current Delta II and Delta III boosters. It comes in a variety of configurations, including a Delta 4M (Medium) (shown here) and a Delta 4H (Heavy). Three Delta 4M rockets were launched between November 2002 and August 2003.

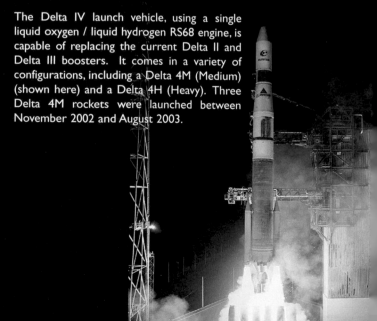

The newest Atlas launch vehicle, based on the same RD-180 engine as the Atlas III, comes in various configurations. Shown here is an Atlas V model 521. The Atlas V is intended for large commercial communications satellites and military payloads. One Atlas V 521 rocket has been launched so far, in July 2003.

Flags of the former Soviet Union (left) and Russia (right).

Russia

The R-7 (below), the world's first intercontinental ballistic missile, provided the basic design for the Sputnik, Vostok, Molniya and Soyuz launch vehicles. The R-7's four-chamber main engines and control engines (a four-chamber engine in the core, and a two-chamber on the side module) can be seen here.

A design common to all of the larger R-7-based launch vehicles, such as the Soyuz booster shown here, was the clustering of four liquid fuel engines into a unit, and then grouping the clusters into a single stage.

The first manned spacecraft was the Vostok ("East"). Shown here preflight is the Vostok in which Yuri Gagarin became the first human being in space. The Vostok launch vehicle was essentially the same design (a modified R-7) that launched the Sputniks. 165 Vostok rockets were launched between Sep. 1958 and Aug. 1991.

The Soyuz is the most reliable and most often used launch vehicle in the world. Its first stage is a direct descendant of the R-7 ballistic missile, and it has a two-engine second stage.

The Cosmos series of boosters were small, two-stage launch vehicles based on the R-12 and R-14 ballistic missiles. They were used in applications where the more powerful R-7-based boosters were not required. The later versions of the Cosmos remained in production until 1995. 608 Kosmos rockets (four models) were launched between October 1961 and September 2003.

The various Soyuz launch vehicle models have been used for both manned and unmanned missions, both military and commercial. 846 production model Soyuz rockets (eight models) were launched between December 1965 and January 2004.

The Proton launch vehicle is a heavy-lift booster that grew out of the UR-500 program, an early contender for the Soviet lunar missions. Capable of delivering 21 tons to low Earth orbit, the Proton is keeping the International Space Station program alive while the US Shuttle is out of operation.

The closed-cycle liquid rocket engine 1D58M, used by the Proton booster, provides 8.5 tons of thrust burning oxygen and hydrocarbon fuel. It was the world's first engine to provide multiple in-flight firing.

The N-1 lunar launch vehicle's first stage was a unique design comprised of 30 engines (nelow) — 24 around the base perimeter and six more in the center.

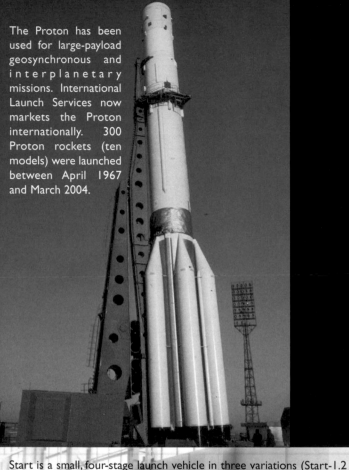

The Proton has been used for large-payload geosynchronous and interplanetary missions. International Launch Services now markets the Proton internationally. 300 Proton rockets (ten models) were launched between April 1967 and March 2004.

Start is a small, four-stage launch vehicle in three variations (Start-1.2 model is shown). Six Start rockets (three models) were launched between March 1993 and February 2001.

The N-1 launch vehicle (above), part of the Soviet lunar program, would have been the world's most powerful booster if it had been successful. The N-1 and entire lunar program were canceled before an actual lunar mission could be undertaken. Failure of any one of its 30 engines, or its supporting systems, would shut down the entire launch. The N-1 was intended to lift a lunar package designated as L-3 (replacing an earlier L-1). Four N-1 (11A52) rockets were launched between February 1969 and November 1972.

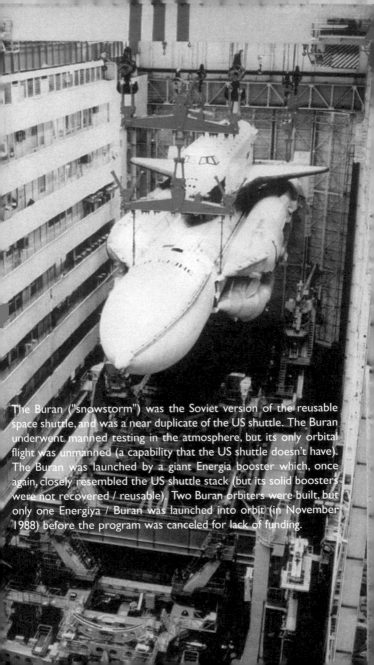

The Buran ("snowstorm") was the Soviet version of the reusable space shuttle, and was a near duplicate of the US shuttle. The Buran underwent manned testing in the atmosphere, but its only orbital flight was unmanned (a capability that the US shuttle doesn't have). The Buran was launched by a giant Energia booster which, once again, closely resembled the US shuttle stack (but its solid boosters were not recovered / reusable). Two Buran orbiters were built, but only one Energiya / Buran was launched into orbit (in November 1988) before the program was canceled for lack of funding.

There were two incarnations of the Energia launch vehicle (also spelled Energiya) – the Buran launcher and a heavy launcher version designated Energia-M, capable of putting 30 tons into low-Earth orbit. One Energiya rocket was launched, in May 1987, after which the program was put on hold due to lack of funding.

China

China's Chang Zheng (Long March) family of rockets has been developing steadily since the late 1960's. China was the third nation to orbit a human being.

The CZ-1 was China's first orbital launch vehicle. It has two liquid propellant stages and a solid third stage, first launched in April 1970. The CZ-2C has a payload capacity of about 2,800 kg. The CZ-2C/SD variation had a "smart dispenser" and an improved second stage. The CZ-2C/SD was used to orbit the Iridium satellites in pairs. The CZ-3 launch vehicle was descended from the CZ-2C, but with a liquid oxygen/liquid hydrogen third stage. It is used for putting satellites into sun-synchronous and geostationary orbits. The CZ-4 launch vehicle is basically a CZ-3 with a different third stage, for putting satellites into sun-synchronous and polar orbits. The The CZ-4B has an improved third stage. The CZ-2E is China's largest launch vehicle to date, with a payload capacity of 9,500 kg, and uses liquid strap-on boosters.

CZ-1   CZ-2C   CZ-3   CZ-4   CZ-2E

The CZ-2D is similar to the CZ-4, with storable propellants, and has a payload capacity of about 3,500 kg for low-Earth orbit satellites.

The CZ-3A is similar to the CZ-3, with an improved third stage, and has a geostationary payload capacity of about 2,700 kg. The CZ-3A is available to commercial customers.

The CZ-3B is descended from the CZ-3A with larger propellant tanks and four strap-on boosters. It has a payload capacity of about 5,000 kg.

The CZ-2F launch vehicle is a CZ-2E that has been upgraded for manned missions, and includes a launch escape tower.

CZ-2D   CZ-3A   CZ-3B   CZ-2E

Japan

Early Japanese launch vehicle designs were based directly on the US Delta series. Continued development has produced uprated designs that can compete effectively with the other space nations. Japan has launched a total of 69 rockets of 14 types. All Japanese launch vehicle photos courtesy of JAXA.

The N-II launch vehicle (above) was a larger version of the N-I, with a geostationary orbit payload capacity of 350 kg. The The N-II had six strap-on boosters (the N-I had two). Eight N-II rockets were launched between February 1981 and February 1987.

The N-I launch vehicle (at left) was a version of the US Delta launch vehicle built under license in Japan. It had a geostationary orbit payload capacity of 130 kg. Seven N-I rockets were launched between September 1975 and September 1982.

The H-I launch vehicle had the same first stage as the N-II, a Japanese-built liquid oxygen / liquid hydrogen second stage, and a solid propellant Japanese third stage. It had a payload capacity of about 1100 kg. Nine H-I rockets were launched between August 1986 and February 1992.

The H-II launch vehicle employs completely Japanese designed propulsion systems, and has a low-Earth orbit payload capacity of 10 tons. The liquid oxygen / liquid hydrogen H-II, with two large solid-propellant strap-on boosters, is much larger than any of its Japanese predecessors. Seven H-II rockets were launched between February 1994 and November 1999.

The J-I launch vehicle is a three-stage solid fuel rocket with a low-Earth orbit payload capacity of about 1,000 kg. The J-I was developed using a combination of existing Japanese rockets. The J-I turned out to be more expensive than comparable vehicles from other countries, so it is currently under review and imported components are being considered. J-I test rocket No. 1 is shown here lifting off from the Osaki Launch Complex.

The H-IIA launch vehicle is an upgraded version of the H-II that exists in models employing var-ious configurations of solid and/or liquid strap-on boosters. Six H-IIA rockets (three models) were launched between Aug. 2001 and Nov. 2003.

This is the V2 rocket. Built during World War 2 by the German government under the leadership of Wernher von Braun. It was the world's first strategic ballistic missile with a single engine generating about 20 tons of thrust. Over 3500 were built. Seen in the background is the Skylab space station, which was a modified third stage from the Saturn V.

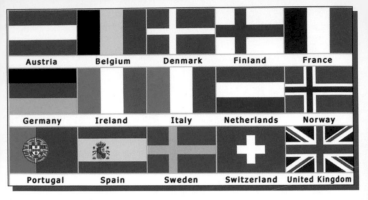

Austria	Belgium	Denmark	Finland	France
Germany	Ireland	Italy	Netherlands	Norway
Portugal	Spain	Sweden	Switzerland	United Kingdom

European Space Agency

Prior to the formation of the European Space Agency (ESA) in 1974 there was a limited amount of launch vehicle development in Europe. The ESA has now had 25 years of success with its Ariane family of launch vehicles, launching a total of 163 rockets of 12 models.

Ariane 1, ESA's first successful commercial launch vehicle, took eight years to develop. It has been used to launch satellites in pairs and the Giotto probe. Eleven Ariane 1 rockets were launched.

Ariane 2 was an upgraded Ariane 1 with engines of greater thrust and larger first and third stages, providing longer burn times and greater payload capacity. Six Ariane 2 rockets were launched between May 1986 and April 1989.

Like the Ariane 2, Ariane 3 was an upgraded Ariane 1. In addition, it had strap-on solid or liquid-propellant boosters for greater payload capacity. 11 Ariane 3 rockets were launched between August 1984 and July 1989.

Ariane-1   Ariane-2   Ariane-3

Ariane 40 42P 42L 44P 44LP 44L

Ariane 4 is a family of six medium / heavy-lift launch vehicles configured by adding solid and/or liquid strap-on boosters. 116 Ariane 4 rockets were launched between June 1988 and February 2003. The Ariane 40 (7 launches) is the basic three-stage liquid-propellant core vehicle.

The Ariane 42P (15 launches) has two solid strap-on boosters. The Ariane 42L (13 launches) has two liquid strap-on boosters. The Ariane 44P (15 launches) has four solid strap-on boosters. The Ariane 44LP (26 launches) has two solid and two liquid strap on boosters. The Ariane 44L (40 launches) has four liquid strap-on boosters.

The largest ESA launch vehicle to date, the Ariane 5 has a liquid hydrogen / liquid oxygen Vulcan main engine and two large solid rocket boosters which give a large part of its take-off thrust.

Ariane-5G

 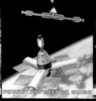